Editorial   Steve Parker

Design   David West
Children's Book Design

Photo researcher   Cecilia Weston-Baker

Consultant   Graham Buxton, Co-Founder and Hon Secretary, Campaign Against Drinking and Driving, UK

© Aladdin Books Ltd 1989

Designed and produced by
Aladdin Books Ltd
70 Old Compton Street
London W1

First published in the
United States in 1989 by
Franklin Watts
387 Park Avenue South
New York NY 10016

ISBN 0 531 10433 8

Library of Congress Catalog
Card Number 89 31584

Printed in Belgium

The publishers wish to point out that all photos appearing in this book were either lent by an agency or shot with posed models.

Contents

Introduction	5
The people involved	7
Laws and attitudes	17
The effects of alcohol	27
The legal limits	37
Penalties for drunk driving	45
Staying safe	51
Factfile	59
Sources of help	60
What the words mean	61
Index	62

UNDERSTANDING DRUGS

DRINKING AND DRIVING

Christine Madsen

FRANKLIN WATTS
New York · London · Toronto · Sydney

INTRODUCTION

When many people are killed or hurt at the same time, in an earthquake or a plane crash, we call it a disaster. There is massive news coverage, and there may be an appeal to raise money for the victims, families and friends, relatives and dependents. When a few people are killed and hurt, the news coverage is much less. When only one person dies, it is rarely national news. There may only be a short announcement on regional radio or a few lines in the local newspaper.

But every individual death or injury is a great tragedy – for the victim, the family and friends, and the relatives and dependents.

Imagine the outcry if more than 120 jumbo-jets crashed at the same time, killing all the 46,000 people on board. Yet the same number of deaths occur in the United States each year, due to road incidents. In Britain, the number of road deaths is equivalent to 15 jumbo-jets. But because they tend to happen in ones or twos, here and there, we rarely hear much about them. Unless one of those killed was someone we knew.

Deaths on the road are often called "accidents." Almost anything we do involves a certain risk of injury, or even death, due to an "accident." Statistics show that driving a car, riding a motorcycle or bicycle, or crossing the road as a pedestrian, all involve a higher-than-average degree of risk.

Wrecking: people injured, cars smashed, travelers delayed.

We might expect that drivers, riders and pedestrians would take extra care to minimize these risks. Yet thousands of people knowingly increase their chances of being injured or killed on the roads, and of injuring or killing someone else. They drink and drive.

Alcohol is directly responsible for thousands of deaths and injuries on the roads each year. Someone who has drunk alcohol has a blurred awareness, poor coordination, slower reactions and mood changes that make an "accident," or crash, many times more likely.

In the United States, 38 percent of drivers and motorcyclists killed in road crashes have drunk more than the legal limit of alcohol. On Friday and Saturday nights in the United States, this percentage becomes even higher.

Why do people drink and drive, when they know it is so dangerous? There is no excuse for drinking and driving. But there are various reasons why people do it. It helps to understand these reasons, and how they are shaped by many factors: our modern lifestyle, our behavior and expectations, the attitudes of people around us, and our laws and social customs. Hopefully, through understanding, we can act to reduce the terrible injuries and waste of life on the roads, by reducing the number of drunk drivers.

> ***Don't drink and drive.*** **Often-used saying to persuade people not to drive while under the influence of alcohol.**

THE PEOPLE INVOLVED

"*We experienced everything when she was killed. We even felt guilty...*"

It is an ordinary night in an average town. Two good friends drive to the local bar for a drink. There, they meet another friend, who has walked to the bar to celebrate her birthday. They each have several drinks. The two friends drive home. On the way, the driver loses control and crashes into another car, and is badly injured. The passenger and the driver of the other car are killed.

A few minutes' thought shows the terrible consequences of this "evening out."

The aftermath

The driver is maimed for life, and will be in a wheelchair. He has caused the deaths of two other people, one of them a good friend. He is alive to reflect on what happened, and why. He has caused suffering to himself, his family and friends, and to the families and friends of his passenger and the other driver. How can he face his dead friend's parents and brother? Some drivers in this situation wish they had been killed instead.

The passenger's family is grief-stricken. They may "blame" the driver, for taking the wheel while drunk. But they also realize that the two friends were "in it together." Why did their son allow his friend to drive, knowing he was drunk? Could he have insisted that no one drive, and called a taxi instead?

The family of the woman driving the other car is devastated. She was driving home from a fitness class, minding her own business, when the fatal crash occurred. Her husband has lost his wife, her children have lost their

The sensible alternative: share a cab to reduce risks.

mother, and her parents have lost their daughter.

The friend who celebrated her birthday is also affected. Should she have encouraged the two friends to stay for extra drinks, knowing they had to drive home? If they had left the bar earlier, with less to drink, maybe they would have been alright. Is she partly responsible for the tragedy?

Feelings and emotions
It is difficult to know in advance how we might react to such a situation. The families and friends of people killed through drinking and driving often report a mixture of feelings and emotions. Anger, resentment, grief, sorrow, regret and guilt are common. They feel so intensely that they are unable to go on with their lives for a while. Many never really get over it.

> **I had never felt such a cauldron of emotions, than I did when my son was killed.** Mother of man killed through driving when drunk.

Some people "bottle up" their emotions, while others let them out with great force. Some of the affected people become withdrawn, refusing to talk about the event or to go out themselves. Others harbor anger and aggression toward anything that brings back memories.

Anger toward everyone
A man was sent to prison after crashing a car. The

Grief and sorrow spill over at a loved one's funeral.

passenger, who was his best friend, was killed in the crash. Both were drunk. In this real-life case, the wife of the driver reacted with anger – at everybody. She was angry at her husband for driving after drinking, at his friend for letting him drive when drunk, at herself for letting them go out. She was even angry at other road-users on that night, thinking that they may have caused him to drive in a certain way, and so be responsible for the crash.

> *I feel so angry at both of them.* **Wife of London man who drove when drunk, and his best friend (also drunk) was killed as a passenger in the car.**

The authorities attend a crash, always alert for alcohol involvement

Sometimes, the death or injury of a drunk driver is seen as "justice." If drivers are drunk, then do they deserve such a "punishment?" But the wife did not see things in such a way. She felt that almost everyone was to blame.

"Murderer"

In one drinking-and-driving incident, the driver who caused the crash had taken so much alcohol he could not stand up. Although his own injuries were minor, he had to be carried to the ambulance because he was too drunk to walk.

The other person, a motorcyclist, was killed. His mother became hysterical when she heard about the death of her only son. For many months she waged a campaign against the driver – but because he was in prison, she acted against his family. She made phone calls, painted "Murderer" on their house, and shouted at the family members in the street. Such are the intensely distressing and tragic effects of drinking and driving, for all those involved.

> **We experienced everything when she was killed. We even felt guilty that somehow we had brought her up wrongly.** Middle-aged Illinois woman whose younger sister was hit by a drunken driver.

The death of friends

Alcohol-related road crashes often lead to the death of a friend of the driver. In fact, the people most likely to die in a drinking-and-driving crash are the driver and the passengers in that car.

In another real-life case, a group of friends went out together for a few drinks. At the end of the evening, the two car owners offered to drive the others home. Nobody was sober. So the passengers did not notice that the drivers were drunk, and neither did the drivers themselves. On the way home, one of the cars crashed. The driver and one passenger were killed, and another passenger was badly injured. A third passenger had only a few minor injuries. His main reaction was guilt, that he "escaped" so lightly while his friends were dead or maimed.

The daily toll

In the United States, it is estimated that more than 23,000 people are killed every year in road incidents where alcohol is involved. More than 18,000 are drivers or pedestrians who have taken more than the legal driving limit for alcohol.

According to the National Highway Traffic Safety Administration (NHTSA), alcohol is a factor in crashes in which almost 560,000 people are injured each year. This averages out to 1,534 people injured every day, or at least one person injured every minute.

Aside from destroyed lives and physical and emotional pain, the cost to society is high in monetary ways too. NHTSA estimates that traffic crashes cost about $74 billion each year for property damage, insurance administration, medical bills, funeral expenses, rehabilitation, and time lost in the workplace.

Driving on today's freeway needs attention and concentration.

The other half

About half of all the road casualty victims are those who drink and then drive. Some victims are in the same car as the alcohol-impaired driver. Many of the others are just people who happened to be in the way. Babies in their baby-carriages have been killed by drivers who could not steer straight enough to keep their cars off the pavement. Bicyclists and pedestrians have been killed by drunken drivers. Other drivers have been killed by people who have lost control of their own vehicle and crashed into them.

Cars driven by drunk drivers have plowed into stores, houses and schools, with terrible loss of life. In one instance, a group of young people on their way back to camp was hit by a van, driven by other youngsters who had been out for the evening on a drinking spree.

Young and old at risk

Death and injury due to drunk driving happen more often to young people than to older ones. In the United States, just over one-quarter of the population is aged between fifteen and twenty-four. But just under half of all people killed on American roads are in this age group. This means people aged fifteen to twenty-four are almost twice as likely to die in a road crash than people in other age groups.

In all American states, the minimum legal age for drinking alcohol was raised from eighteen to twenty-one years old. The numbers of people dying in highway crashes fell, on average, by one-tenth after this change. Several other states are now considering making the same changes.

We drive to the store to buy alcohol, but not under its influence.

Action in the aftermath

Some people who have lost a loved one due to drunk driving feel that they must become involved somehow, in telling people of its risks, and the enormous and tragic loss of life. They feel they must take action. They may work in a small way to publicize the dangers, such as talking to and persuading others in everyday conversation. Some campaign in a more open way, appearing on radio and television and writing in newspapers, books and magazines.

To people who have not been affected by drinking and driving, these campaigns may sometimes seem to be rather emotional. But think for a few moments about how death or injury to a loved one might affect you. It is not hard to imagine how deeply we might feel about something – or someone – that hurts or kills the people we love.

> *Stay low.* Advertising campaign to encourage people not to drink and drive at Christmas. It met with mixed success.

LAWS AND ATTITUDES

"*If I was sober, it probably still would have happened.*"

Our attitude toward drinking and driving is affected by many aspects of everyday life. We are ruled by social customs and laws, and the way we behave is also guided by our personal pride and sense of importance. The way other people behave can also have a powerful effect on our own actions. Sometimes we try hard to be like a person we admire, and sometimes we simply want to rebel and be different.

For certain people, driving a car or motorcycle fast is a way of showing off and telling the world how fearless and brave they are. But truly skillful driving is not just a matter of going fast. It involves driving safely, with consideration for other road users.

Drinking alcohol is in some ways similar. Certain people seem to think that going out and having too much to drink makes them look good. Others view this as immature and foolish. They see "skillful drinking" as knowing how much to drink under the circumstances – with no alcohol at all if driving.

Feelings about "accidental crashes"

For most people, the way they feel about a road crash depends on various factors. If no one is hurt and nothing important is damaged, the event may hardly be noticed. But a serious crash, where a person is injured, is more likely to make us think about what happened, and why. We may feel that we want to "blame" somebody. This may be partly because, if we know why it occurred, then we can act to prevent it happening again or to reduce the risks of

The head-on collision, where combined speeds exceed 100mph.

recurrence in the future.

> *The road was slippery. If I was sober, it probably still would have happened.* **Driver who crashed the car into a ditch after drinking.**

However, we are not always sensible and objective about finding the cause of the crash. It usually seems easier to find fault with the "other person" rather than with ourselves. It also usually seems easier to blame people who are not like us. Suppose a bicycle and a car are in collision, and you are a keen bicyclist. When you hear the news, you may well be more sympathetic toward the other bicyclist rather than toward the driver of the car. In other words, you "identify" with bicyclists.

Drivers tend to identify with other drivers, even ones who have had crashes (although not crashes linked to alcohol). They may be able to imagine that the same thing could happen to them one day. "Accidents happen," and no driver has magical protection.

> *It wasn't my fault, it was an accident. And we were all drunk.* **New York driver who smashed into a traffic sign while drunk.**

Feelings about bad driving

However, suppose we are not talking about "accidental crashes." Suppose we consider crashes due to deliberately bad driving. Would other drivers identify with, and feel sympathetic towards, a dangerous driver?

Probably not. Few drivers believe that they themselves drive badly or dangerously. So it is doubtful that many drivers identify with other, really bad drivers.

In a similar way, most drivers do not identify with drinking and driving. A survey in Britain showed that about four out of five people believe that it would be a good idea if any driver could be stopped at any time for a breath test, to see if he or she had been drinking. (This is the "sobriety checkpoint.") Not all of these people were drivers. Yet in a more recent survey, three out of four drivers support random breath testing – and these were drivers who admitted to drinking and driving themselves.

Restricting personal freedom?

Some people still feel that sobriety checkpoints would restrict individual liberty. They believe that, provided people do not have a crash or drive dangerously enough to be noticed, their lives should not be interfered with. Yet we all have to give up some personal freedom, to be able to live together in society. We have laws, rules and regulations over many parts of our lives.

Who would "suffer" if the police could stop any driver at any time, to see if he or she had been drinking too much alcohol to drive safely? For the sober driver, it might mean the journey took a few minutes extra. For the driver who has drunk too much, it could mean a whole lot more – but then, he or she is breaking the law, and is a potential danger to other road users. Not only that, drunk drivers are a danger to themselves. In the United States, they make up about 40 percent of the people killed in alcohol-related road crashes. Would a random breath test law act as a deterrent and discourage people from drinking and driving? In New South Wales, Australia, random breath testing has been introduced. The police believe that this is the reason why about one-third fewer drunk drivers are killing themselves on the roads.

Obeying the law

We know that drinking and driving make a lethal combination. So why do people still do it? Why do we need laws to make us do what we should do voluntarily?

> *It took the injury of a close friend to stop me drinking and driving. If that hadn't happened, I might still be doing it.* **Woman who depended on her car for her job.**

In many countries, cars must have seat belts, but wearing them was not universally compulsory. In recent years, various countries have introduced laws to say that drivers and front-seat passengers must wear seat belts.

In the United States, injuries on the roads had increased every year since record keeping began. In one year, seat belt wearing was made compulsory in one state. This was the

The sensible start to a drive – belt up, and stay sober.

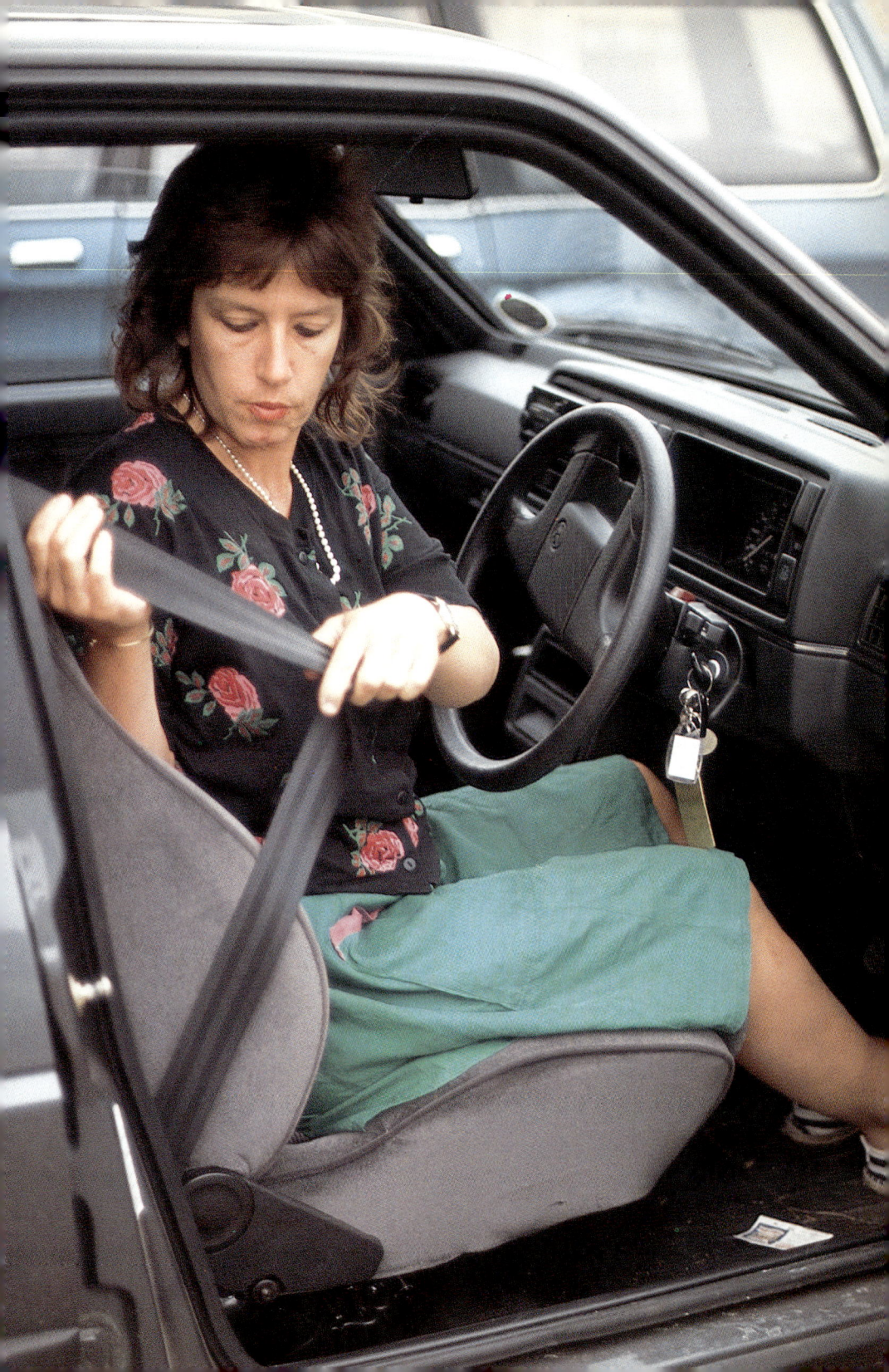

only time when injuries stayed at the same level as the year before.

The seat belt laws were passed in Britain in 1983. A big advertising campaign was conducted to persuade people to use seat belts, by showing how they saved lives and reduced injuries. Before the laws, only one person in five used a seat belt. Soon after the laws were introduced, 19 out of 20 people wore them. In the following year, about 1,000 fewer people were injured in road accidents. Several years later, most people see "belting up" as a sensible, safe and natural part of driving.

Changing attitudes

A real-life case shows how, over recent years, attitudes have slowly changed with respect to drinking and driving.

Four friends used to take turns to drive them all to the sports club one evening each week. After their exercise, and naturally thirsty, they used to have a few alcoholic drinks at the bar in the sports club. But occasionally a "few drinks" turned into "a few more" drinks. On the way home, whoever was driving had usually drunk too much alcohol to be safe.

Then another member of the club was stopped by the police and found to be "over the limit." He lost his driving license. Since he needed the license for his job, he had to find other work, which was less well-paid.

The four friends decided it would be more sensible that whoever drove would not drink any alcohol at the sports club bar. Then they would drive to a bar nearer their homes. Here, the driver could have a few drinks, and he would only have a

In country areas, plan any drinks so that someone can drive safely.

short distance to travel to his house.

A year later, there was a fatal road crash in the area. The dead driver had been drinking. The four friends did not know anyone involved, but it made them think again about driving after even a small amount to drink.

They decided that they would have one drink at the sports club, then drive home and leave the car, and walk to the local bar from there.

Gradually, their routine has changed. The changes came about partly as a result of knowing about deaths and injuries from drunk driving. They were partly a result of their knowing about the punishments which the authorities can impose. And finally they came to realize that drinking and driving is stupid and dangerous. Will they make more alterations to their routine in the future?

Are things changing?

It seems that some people find it hard to change their behavior, without having the authorities checking up on them or passing laws to force the changes.

But attitudes and behavior are changing, slowly and surely. The number of people driving after drinking more than the legal limit of alcohol has gone down since 1973 – by about one-quarter in Britain, and by more in the United States.

> **My son persuaded me to stop drinking and driving. Not by saying anything, but just by being there and needing me.** New father from New Zealand with three-month-old son.

THE EFFECTS OF ALCOHOL

"I can drive better after a couple of drinks."

Alcohol is common as a social "loosener", to help people relax.

For some people, drinking alcohol is mainly a social habit. It is part of our way of life, something we do now and then, for example on "days off," when we are on vacation, or among friends. Alcohol makes most of us feel more relaxed, less tense and self-conscious, and allows us to relate more freely to one another.

However, alcohol is a drug. It can damage various parts of the body, such as the brain and liver. Taken regularly, it can be addictive. The alcoholic must have a lot to drink every day, just in order to function "normally." This can kill, slowly. And too much alcohol at any one time is dangerous.

Alcohol as a "pick-me-up"

When we have a drink or two, we "loosen up." This is

because alcohol helps us to feel less inhibited. It becomes easier to talk to people we do not know, because we feel more confident, alert and aware. But, although alcohol makes us feel more stimulated, in reality it has the opposite effect. This drug is not a stimulant, but a depressant. It works by temporarily "depressing" (slowing down and disabling) the brain. Its effects depend on which part of the brain is depressed, and what that part does.

Different parts of the brain are depressed by different amounts of alcohol. After a small amount, the first part to be affected is the part that keeps us "under control," the inhibiting part. With inhibitions gone, we feel a sense of stimulation. Some people seem to change their character when they drink, because their self-control fades. Shy

"Larger-than-life" behavior, as inhibitions disappear.

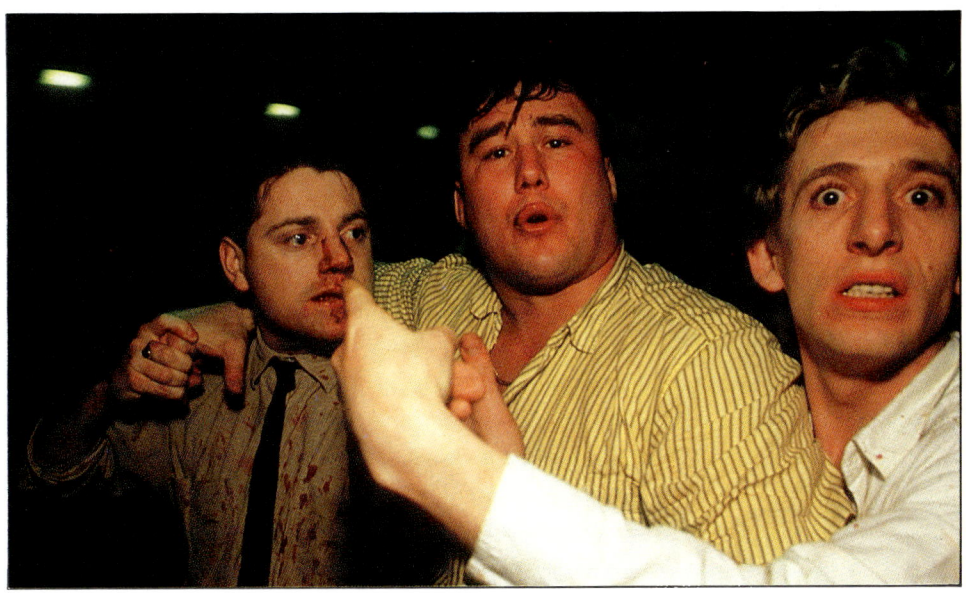

people can become outgoing, and normally gentle people may even become violent, because they have lost their inhibitions.

Altered awareness

After loss of inhibitions, more alcohol starts to depress other parts of the brain and nervous system. This affects our ability to speak, see, move – and think. Speech slurs and vision blurs. We find it difficult to keep our balance and make accurate movements. Our reaction times slow down (page 33). The overall result is that we have clumsy movements and muddled thinking. To increase the problem, alcohol dulls our awareness and alertness, so that we do not realize how badly we are affected.

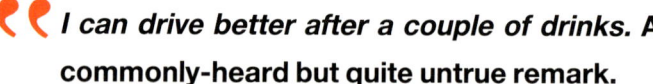

I can drive better after a couple of drinks. A commonly-heard but quite untrue remark.

False self-confidence

Many tests on drinking and driving have been carried out on professionals who drive all sorts of vehicles, from racing and rally cars to buses and trucks. The tests showed that alcohol interfered with driving skill. It made the drivers feel as though they were driving better and making more accurate judgments. But the reverse was true. The drivers did not realize it, even though they were experts, who had good skills and were supposed to be aware of the risks.

In one test, they were asked to drive at speed in and out of a row of traffic cones, without touching them. Normally,

Loss of coordination leads to loss of the drink that caused it!

these drivers would knock down hardly any cones. After one drink, all the drivers thought they were doing well, but all knocked over some cones. After more alcohol, they all thought they could do the test again even better – but they did worse.

> **❝ *I thought that the test became considerably easier. But my score was worse.* A truck driver taking part in a drinking-and-driving experiment in Berkshire, England. ❞**

Another test was carried out with experienced bus-drivers. After just a single whiskey, three of them thought that they could drive their bus through a gap which was

Stop and think. Consider if it is safe to drive.

actually narrower than the bus, by 14 inches (35 cms).

False feelings of safety

One reason why alcohol makes people overconfident is that it takes away the normal response to danger. Instead of becoming alert to a threat, such as nearby traffic, alcohol tends to make us feel artificially safe. It is as though we do not really believe what we see. People seem courageous and brave after they have had a few drinks, whereas in fact they cannot recognize the danger.

Drivers are not the only ones who are at risk from this problem. A study in Glasgow, Scotland, showed that a large proportion of pedestrians injured on the roads there were themselves drunk. Presumably they did not think the moving traffic was a threat to their safety. It is also likely that they misjudged the speed of the traffic, or could not move out of the way quickly enough because of slow reactions.

Slowed reaction times

Another dangerous effect of drinking alcohol is the way it affects the ability to become aware of something and then react quickly to it. Driving is a tricky job. It involves guiding a car or motorcycle along the road, avoiding other cars, reading road signs, and watching for people about to step into the road or for unexpected obstacles. It needs a clear head and good reactions.

Alcohol slows the brain. Somebody who has been drinking takes much longer between seeing a problem and reacting to it, for example, by putting on the brakes. He or she may do the right thing – but too late.

In the hands of the drunken driver, a car becomes a murder weapon. **Slogan of campaign to stop drinking and driving.**

Alcohol in the body

Alcohol gets to the brain very quickly. As soon as it is swallowed, it goes into the stomach and intestine, and from there it is absorbed into the bloodstream. Within minutes, it reaches the brain. In general, the level of alcohol in the blood reflects the general level in the body. The level is affected by factors such as age, sex, body size (page 40), and what else is in the stomach at the time. Alcohol gets to the brain more

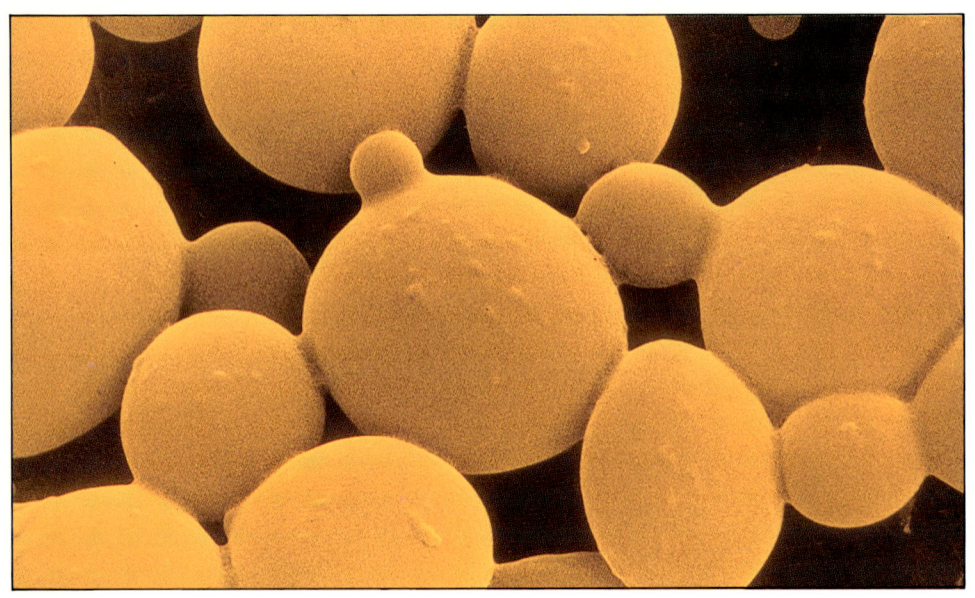

Microscopic yeast produce alcohol as a by-product in fermentation.

quickly through an empty stomach, and fizzy drinks also speed its action.

Getting rid of alcohol

Once in the body, it takes time for the alcohol and its effects to "wear off." The drug is mainly broken down into harmless substances by the liver. Small amounts are given off in breath (which is how the breath test works), in sweat and also in urine (which is why urine tests are sometimes used).

On average, the time it takes to get rid of the effects of one unit of alcohol (see panel) is about one hour. This means that if an average adult drinks an average bottle of beer, the body will have absorbed it and then "used it up" after about two hours.

The alcohol from successive drinks adds together. If a person drinks a bottle of beer at nine o'clock, and another at half past nine, then by eleven o'clock, the alcohol from the second bottle is still in the body, as if it had been swallowed only ten minutes earlier. Suppose somebody drinks as much as eight or nine standard bottles of beer (the equivalent of two bottles of wine) in one evening. There may be the equivalent of two or three bottles still there at seven next morning, when he or she gets up and drives to work.

Units of alcohol

The concentration of alcohol varies in different drinks, and the amount in a "serving" of drink varies from place to place. Therefore alcohol is often measured in *units*. In general, the following drinks each contain one unit of alcohol:

one glass of beer, or hard cider

about three-quarters of one standard can (560 mls) of the same

a sherry-glass of sherry

a "shot" of hard liquor such as whiskey or gin

How the risks increase

Doctors warn that, on average, health is impaired by a regular intake of more than 21 units of alcohol per week for men, and more than 14 units per week for women. Even below these levels, damage can occur to the body.

35

After one unit of alcohol (slightly less than a bottle of beer, or a glass of wine, or a shot of hard liquor), the chances of having an accident is two times normal. After two units of alcohol, the chance of having an accident is three times normal. After four to five units of alcohol (the legal limit in some places), the chance of an accident is five times normal.

"Experience" in drinking
Young people are more affected by alcohol than older people. Alcohol is a drug and, like other drugs, the body can get used to it. A person who has been drinking alcohol regularly for many years can take in more before seeming to become affected. However, this amount of drinking is almost certainly damaging various parts of the body.

The ability to drive also changes with experience. A new driver needs more time to think about what to do, than somebody who has been driving for some years.

The statistics combine in the following way. A new, young driver is more likely to have a crash than an experienced, older one. A young drinker is more likely to be affected by alcohol than an experienced, older one. The risks add together. The result: almost half of all people killed in road accidents are under 24 years old.

> **In the dream, I crash into my own car.** Drunk driver from California who injured his wife and became mentally disturbed.

THE LEGAL LIMITS

"I told the police I'd had two beers..."

In times past, if the authorities suspected a driver was dangerously intoxicated, they had to use their own judgment. For example, the police might stop a suspect driver and listen to how he or she spoke or answered questions. The driver might be asked to walk in a straight line, to test balance and coordination.

Nowadays, in most regions the police can ask a suspect to take a chemical test, by providing a sample of blood, breath or urine. This is analyzed to find out how much alcohol it contained. In Britain, the breath-test device or "Breathalyzer" was introduced in 1967. It can be used as a roadside test to indicate more than a certain level of alcohol in breath. If it is positive, the driver may be asked to give a blood sample, for a blood test. (The exact procedure varies from place to place.)

> *I told the police I'd had two beers, hoping they would believe me. The blood test showed six. Actually I can't remember.* **Driver who was stopped because his car had a faulty headlight, and who was then breath-tested.**

The "limits"

Most countries or states have set a "legal limit" — the level of alcohol in the blood taken as absolute proof that a driver is impaired through alcohol.

In Britain and several other countries, and in Idaho and Utah in the United States, the "legal limit" is 80 mgs

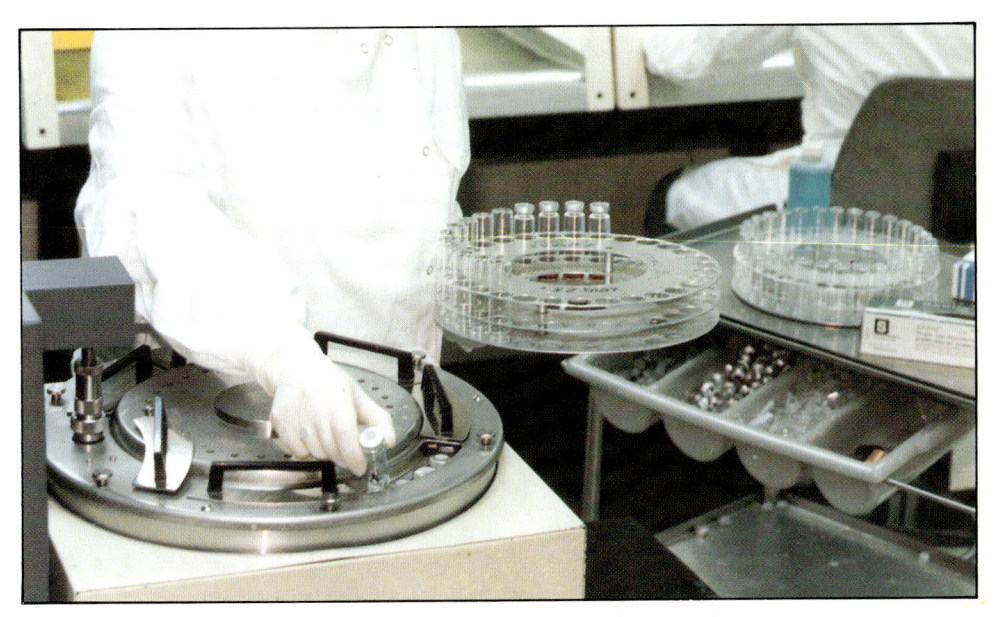

Analysis of samples of blood or urine is fast and accurate.

(milligrams) of alcohol in 100 mls (millilitres) of blood. This is often written as "80mg/100ml". On the breathalyzer scale, alcohol is measured in µgs (micrograms), and the limit is 35 µgs of alcohol per 100 mls of breath.

How much alcohol is this? In general, it is the amount contained in about three to five drink units – say, three 16oz bottles of beer or four glasses of wine. But this is only an average. The actual amount of alcohol in somebody's blood *cannot* be predicted accurately from what he or she has drunk. It depends on many factors, such as age, sex, body weight, whether there is food in the stomach, and even if it is a hot day and the person has been perspiring a lot.

Some countries, such as Canada, have two "limits." People over the higher level may be punished more severely.

than those over the lower one.

State or country	Legal limits
About four out of five US states	1.5 grains (100mgs) alcohol in 2.7 fluid drams (100mls) blood
Britain, Canada, the states of Idaho and Utah, New South Wales, Australia, several European countries	1.2 grains (80mgs) alcohol in 2.7 fluid drams (100mls) blood
Japan, Greece, Holland, Poland, Yugoslavia, Victoria and Western Australia, Australia, most Scandinavian countries except Denmark	0.75 grains (50mgs) alcohol in 2.7 fluid drams (100mls) blood
Sweden	Any level above the normal "background" body level
Most other states in the United States	No legal limit specified as yet

Which is the correct limit?

It is clear from the varying figures that the "safe" amount of alcohol for drivers has yet to be agreed. This is not surprising, however, since these are relatively recent laws. Research is still being done all over the world, to see in more detail how alcohol affects people's ability to drive, and to assess the effects of the present laws.

Before you begin to feel the effects of your first drink, you are twice as accident-prone as if you had no alcohol. By the time you start to relax, you are already three times as accident-prone. If you go to the very brink of the United Kingdom legal limit you are five times as accident-prone. Staying just under the limit may mean you would not be prosecuted for being legally too drunk to drive. But it does not mean you can drive safely, and it will not protect you from the high risks of being in a crash.

More than 20 years have passed since some laws were introduced. A whole generation of people have grown up who have heard all about the dangers before they ever start driving or drinking. Public opinion is changing, year by year. Many people feel the limit should now be lowered. An increasing number believe that it should be against the law to drive after drinking any alcohol at all.

> *I never do it. Most of my friends don't, either. The older generation do.* **19-year-old driver from New York.**

After the crash, talk about "limits" takes on new meaning.

Drinking "up to the limit"

Without a Breathalyzer or blood test, nobody can tell when they go over the limit. Generally speaking, small people use up alcohol more slowly than big people; thin people feel the effects more quickly than fat people; and a woman's chemistry and body tissues deal with alcohol more slowly than a man's. Add to these the variations caused by how much food a person has eaten, of which kind, and even the stage of the menstrual cycle for a woman. It is clear that trying to "stay just under the limit" is far too risky.

> *I thought I'd be under the limit.* Often-heard excuse from over-the-limit drivers.

About one-third of drivers who are tested on the Breathalyzer after crashes or other incidents turn out to be over the legal limit for alcohol. In addition to the injuries they have caused or suffered, and damage to vehicles and property, there will be the inevitable penalties for drinking and driving.

A routine scene at incidents, checking for presence of alcohol.

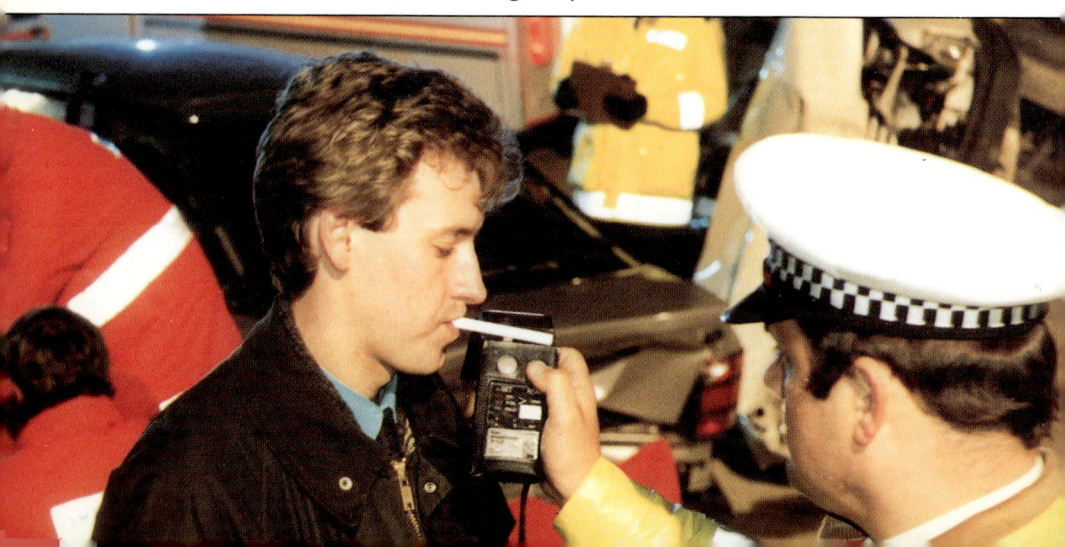

Education at a younger age

In the United States in 1988, the National Commission Against Drunk Driving (NCADD) released a report on Youth Driving Without Impairment. It was compiled from information and interviews conducted in five centers: Atlanta, Boston, Chicago, Fort Worth and Seattle.

The report confirmed that alcohol-related road incidents remained the leading cause of death for young American drivers, and it made wide-ranging recommendations.

> *We must end the senseless behavior that makes alcohol-related crashes the American way to die for youth of driving age.* **Youth Driving Without Impairment report, NCADD, 1988.**

The NCADD report concluded that drunk driving among the young was a general social problem, and it required coordinated action from a number of directions. These included school-based and work-based information programs, campaigns centered in the local community, driving license and drinking age changes, and appraisals of how the laws were enforced and of the laws themselves.

Talking through the issue with the young people concerned led to some interesting results. In a study at Michigan State University, high school students were asked which factors would reduce the amount of alcohol they drank. Between 70 and 90 percent said that keeping tighter control over the household's alcoholic drinks, taking more interest in their social life (especially at weekends), and supervising at gatherings and parties would help.

The study also revealed that two parents out of three believed that drinking by under-aged young people happened, but only one parent in five believed that their own child would be involved.

There was also dissatisfaction with the way that the regulations were sometimes enforced. An older person arrested for drunken driving is often subject to the full weight of the law. Yet for young people in the same situation, the offense has sometimes been "down-graded" and made to seem less serious. For example, the guilty young person might be entered into a pre-adjudication program or even simply given over to the care of the parents.

> **The key is for communities to put all of these elements in place, so that the efforts of students, parents, schools, courts, businesses and police support one another.
> Youth Driving Without Impairment report, NCADD, 1988.**

PENALTIES FOR DRUNK-DRIVING

" A driver may be forbidden to drive for life. "

In the United Kingdom, if someone is hurt in a car crash, the police must be told. It is illegal not to report an accident where a person has been injured. In most states in the United States, the police can also ask a driver to take a breath test if they suspect he or she has been drinking because the person is driving oddly or badly. Or it might happen that the driver is stopped for another reason, and the police then notice he or she seems drunk.

Being "over the limit"
If a breath test is positive, the person involved will be asked to go along to the police station, to take another breath test or possibly give a blood sample. In many places in the United States, refusing to do this is regarded as presumptive evidence of guilt, and may result in loss of license rights. In Britain, someone who refuses to take a breath test is fined, and refusing to provide a blood sample brings a bigger fine. Depending on the area, the over-the-limit driver may then be summoned to appear in court.

Excuses are hard to find
In the early days of breath testing, people thought up all sorts of excuses. These occasionally worked, because the laws were new and unfamiliar. Today, the laws are more thorough, and the authorities are more aware about people trying to find "loopholes".

Tricks developed in the past such as having an on-the-spot drink before being tested, or running away from the scene, or driving home quickly and getting out of the car, no longer work. When a test is taken, it is even possible to calculate backward and find out what the alcohol level was at the time of the crash.

> *He panicked and ran off across the field, but they soon caught him. It only made things worse.* **Friend of man convicted of drinking and driving.**

The legal penalties

Punishments under the laws for "driving while intoxicated" vary from area to area, but generally they are severe. In the United States, being convicted for driving "over the limit" means an automatic suspension of drivers' license for at least one year, and possibly longer. Even for a first time conviction, there may be a large fine and perhaps a

The sworn testimony in court is only one part of the evidence.

six months' prison sentence. Being found "in charge" of a motor vehicle while over the limit (though not actually driving it) can bring a six-month driving ban and/or a $1,000 fine.

In the United Kingdom, people can be disqualified from driving for much longer than one year, if very drunk or if guilty of the same offense before. After three or more offenses, a driver may be forbidden to drive for life.

In the United States, a second conviction usually is a felony, punishable by longer prison sentences, heavier fines, and suspension of driver's license, often for life.

Subsequent problems

After the legal penalty has been completed, the driver is still not back at the start. In Britain, the driving license will be "endorsed" with penalty points: 4 after a ban, but 10 if he or she had only a fine. When a driver has collected 12 penalty points, a driving ban is compulsory the next time he or she commits a traffic offense.

The convicted drunken driver has other problems to face. Car insurance becomes more expensive, and some guilty drink-drivers find it hard to get any insurance at all. If driving is important to that person's job, he or she may not be able to continue with their employment.

Punishment by law is not the only punishment an over-the-limit driver faces, even if no one was injured or killed. Family, friends and neighbors can take a dim view. Most people now see drinking and driving as irresponsible and downright dangerous, and it certainly does not inspire respect. Having a conviction for drinking and driving can affect many areas of life, such as applying for a new job, for elected office, or financial credit.

> *When he came out of jail, he got no pity. We ignored him for a long time. He really paid for it.* Friend of Chicago man imprisoned for crashing into another car when drunk and injuring the driver.

Consequences of injury or death

All the above can happen when no damage has been done. Damage to property, or injury, brings more legal charges. Any driver, over the limit or not, who hurts somebody in a crash may be charged with careless or reckless driving. If it comes to a legal battle between a sober driver and one who

In many countries around the world, car crashes occur daily.

has been drinking, the drinker is unlikely to succeed – even if the other driver could have been at fault.

Drivers who injure or kill someone have to live with their feelings for the rest of their life. This is a terrible burden, even for a driver who was honestly not to blame for a crash. If a driver knows the accident happened because of alcohol, the burden can be much, much worse, knowing that someone's life has been needlessly ruined, or even ended.

> *I'll blame myself for the rest of my life, but I thank God I hadn't been drinking.* **Driver who hit a child that ran into the road without warning.**

STAYING SAFE

"Your best friend is the one who won't buy you a drink if you are driving."

No one has to drink and drive.

After a road crash or being found "over the limit", most people who drink and drive feel the same way: "If only..." If only they had not been drinking – or driving. If only they had not had that last drink. If only they had traveled on a bus or train. If only they had known, before the crash, how they would feel afterwards. If only...

Those drivers learn from experience. But others do not have to learn the hard way. This applies especially to young people who have not yet got into the habit of drinking, or driving, or both. You can learn from other people's mistakes.

Sometimes, it is easier to make the decision than it is to carry it out. It sounds simple enough. Don't drink. Or don't drive. But how easy is it, really? It can be easy – if you start in the right way, before getting into bad habits.

❝I've never had a drink when I've been driving. If I want to go drinking, I don't drive. It's easy, if you never start. **21-year-old delivery driver from Washington D.C.**❞

Doing without alcohol

It is perfectly possible to go to a party and have a good time without drinking alcohol. Many bars and clubs serve non-alcohol beers, lagers and wines. These are a useful alternative to the alcohol versions, especially if you are dancing or talking a lot and get thirsty. Often, the general atmosphere and "party spirit" will carry you with it, without the need for alcohol.

Provide alcohol-free drink at parties, for safety's sake.

If you arrange a party, be sensible and provide plenty of non-alcohol drinks such as fruit juices, fizzy drinks, and non-alcohol beers and wines. If you go to a party or bar and these types of drinks are not on show, ask for them. If they are not available, ask why. It should not be a problem to take the sensible and legal alternative. This applies to drivers and also anyone else who wants to have a drink, but not with alcohol in it.

Having just one drink
Some people choose to stop drinking alcohol after just one glass. Surely, one small drink does little harm? This might be so – except for the way that alcohol removes our inhibitions. One drink may be all that is needed to make a second one

seem like a good idea. This is the difficult part. It may be easier to decide not to drink at all. But it is usually much harder to stop after a few drinks. The alcohol begins to take over, and decisions become distorted.

> **We thought he'd had a bit, but we thought that you don't interfere with a mate. Then we thought, yes you do, that's what mates are for.** 25-year-old beauty consultant, London.

The safest way out of this problem is to avoid alcoholic drinks from the start. We can all help, by being responsible. The person giving the party, or the owner of the bar, is not much of a host if he or she encourages everyone to drink

The sensible (and healthy) alternative: juice instead of wine.

alcohol – including drivers. Even with a poor host, there is always water in the tap.

Pressure to drink

Young people learning to drive are under a lot of pressure to drink alcohol. It can come from many sources, such as advertising, friends, family, and from those who believe what the advertisers want them to believe. But young people have advantages their parents did not have. The dangers of drinking and driving are now well known. Breweries make low-alcohol and alcohol-free drinks. Many clubs have started to serve coffee or tea.

Breweries and bar-owners may say that they are leading the way in reducing the terrible consequences of drinking

The brewer's vat, as yeast ferments and makes alcohol.

and driving. However, it seems that some of them are being forced to change, because the way people in general feel about drink is changing. Not long ago, it was hard to refuse a drink without looking weak and wimpish. Now it is perfectly normal for anyone – not only drivers – to say "no" to alcohol.

> **I got into the habit of saying: 'No thanks, I'm driving.' Some of them used to sneer, but they don't now.** 19-year-old keen party-goer, Vermont.

We can all help. Do not force alcoholic drinks on someone you know is driving. Discourage your friends from doing the same thing. This can be difficult, especially when everyone has had a drink or two and their inhibitions are gradually loosening up. If you see someone drinking and you know they will soon be driving, it might help to point out the dangers in a sensible way.

> **Your best friend is the one who won't buy you a drink if you are driving.** Warning displayed in pubs, clubs and bars.

Doing without driving

The safest alternative to going without a drink is to go without driving. There are other ways of getting around, such as buses, trains, and taxis. In a city, it is often easier to use public transportation than to drive, especially if parking is a problem. Remember that driving anywhere costs money for

A safe journey in prospect, having not drunk alcohol.

fuel. A shared taxi fare may not work out too expensive when shared among friends.

Public transportation is not available everywhere, and some places are too far away to walk. One sensible solution is to take turns driving. Friends help one another in all kinds of ways, and this can be one of them. When there are two or more drivers in a group, it makes sense to work out who should stay clear of alcohol and drive home. In return for driving friends home safely, the driver should expect the friends to help him or her not drink – and still have a good time. Making fun of a person who is not drinking in order to drive for others is extremely unhelpful. Another possibility is to ask someone to pick you up afterward – a friend, parent or neighbor. Sometimes staying put, rather than going home, can be the wisest move. The telephone can be used to inform a parent or relative that you are spending the night at a friend's house, or to ask for a ride.

Coping with the unexpected

Plans are sometimes wrecked by the unexpected. A girl went out in her car for the evening with her boyfriend. He did not drink alcohol because they planned that he would drive home. But they had an argument. He walked out and telephoned a friend to take him home. After an argument and four drinks, she was angry and drove home too fast. She exceeded the speed limit, was spotted and stopped by a police car – and the rest you can imagine. She had no problem worrying about drinking and driving for the next two years, since she was banned from driving for this period.

It is not unknown for practical jokers to "spike" other people's soft drinks. Doing this to a driver's drinks is the peak of stupidity, and can end in death.

Trust to luck?

What are the chances of "trusting to luck" when drinking and driving, and gambling that you will not be caught? Ask the friends and relatives of the people who are killed every day by drunken drivers. Ask those who have been injured, or their friends and families. They would have an answer.

Nobody ever has to drive if they do not want to. And if you have been drinking, you do not want to.

FACTFILE

Units of alcohol
A bottle of ordinary beer, a glass of wine, a single shot of liquor, or a small glass of sherry each contain about 1 unit of alcohol.

A bottle of ordinary hard cider contains about 1.5 units of alcohol.

One large can of ordinary beer, a bottle of strong beer, a double shot of liquor or a bottle of strong cider each contain about 2 units of alcohol.

Alcohol removal from the body
One unit of alcohol takes at least one hour to be processed by the body. Only one unit at a time can be processed.

The brewery product: handle with care.

Equivalent levels
A level of 80 mgs of alcohol in 100 mls of blood, is equivalent to a level of 107 mgs of alcohol in 100 mls of urine, or 35 μgs of alcohol in 100 mls of breath.

Effects of alcohol on driving ability
● Reduced muscular control and coordination.
● Blurred and less sensitive vision.
● Impaired judgment of speed and distance.
● Impaired ability to make decisions, especially quickly.

Precautions
If you must drink, then do not drive. Otherwise, help yourself by following these precautions when you drink:
● Do not drink on an empty stomach.

Alcohol belongs on the roadside, not on the road itself.

● Drink slowly.
● Eat at the same time as drinking.
● Limit your intake to 3 units only.
● Stop drinking at least half an hour before you intend to travel.

Penalties for drunk-driving
The legal penalties include loss of license, endorsement of license with penalty points, fine of up to $2,000, prison sentence.

There may be other, more long-term consequences. For example, a conviction or driving ban may result in the loss of your job, and restrict your social life.

SOURCES OF HELP

Many organizations are involved with the various aspects of drink-driving, such as the devastating effects on victims and the convicted, the medical and other effects of alcohol, general aspects of road safety, and the campaign to raise awareness and reduce the numbers of people who drink and drive. If you wish to learn more, in the first instance, write to the address which seems most appropriate.

Help and support in stopping drinking
*Alcoholics Anonymous
PO Box 459
Grand Central Station
New York, NY 10017
(212) 696 1100
   Telephone 01-222 3454*
AA is a self-help group founded by and for alcoholics. You do not have to quit drinking for any length of time before you can participate, the only requirement is the desire to stop drinking. Local chapters are listed in the phone book.

*National Self Help Clearinghouse
33 W 42nd Street
New York, NY 10036
(212) 840 1259*
Can provide information on self-help rehabilitation organizations in your area, or put you in touch with one of the twenty-seven state and local self-help clearinghouses around the country.

*National Highway Traffic Safety Administration (NHTSA)
400 Seventh Street, S.W.
Room 5130
Washington, DC 20590*
The NHTSA are involved in road safety, and record accident statistics. They are also concerned with the legal aspects of road safety.

For family help
*Al-Anon Family Groups
PO Box 862
Midtown Station
New York, NY 10018*
A support group for friends and relatives of people with drinking problems. Al-Alteen is a service of Al-Anon specially for younger family members.

Referrals, Information and More
*Women's Alcoholism Center
2261 Bryant Street
San Francisco, CA 94110
(415) 282 8900*
This center has both a residential and an outpatient facility for alcoholic women and their children. Provides individual and group therapy (including psychotherapy and play therapy) for children, and offers alcohol and drug education.

*National Association on Drug Abuse Problems
355 Lexington Avenue
New York, NY 10017
(212) 986 1170*
Conducts a drug prevention program and offers family counseling.

*Helping Youth Decide
National Association of State Boards of Education
PO Box 1176
Alexandria, Va 22313
(703) 684 4000*
Write for their free booklet about making informed decisions concerning alcohol, smoking, drugs and other issues. This organization also organizes parent-student workshops and community projects.

WHAT THE WORDS MEAN

alcohol the drug, ethyl alcohol (ethanol), contained in intoxicating drinks or liquors

blood alcohol the level of alcohol in a person's blood, measured in milligrams of alcohol per 100 millilitres of blood (mg/100ml)

breath alcohol the level of alcohol in a person's breath, measured in micrograms (μg) of alcohol per 100 millilitres of breath

breathalyzer the equipment for carrying out a breath test. The person breathes steadily into a mouthpiece and the device automatically registers the alcohol content of the breath sample

blood test a test carried out by a doctor or nurse, to take a sample of blood for analysis, to find out how much alcohol is present

drink driving driving a vehicle while impaired through the effects of alcohol

drunk having one's senses, actions and judgement impaired by alcohol

Dutch courage false confidence that comes from alcohol dulling the brain so that it fails to recognize genuine danger

intoximeter a type of breathalyzer

over the limit having a blood alcohol level that exceeds the legally permitted level for a driver in that country or state

penalty point a way of marking the driving license of a person who has committed a driving offense. In the UK, penalty points add up, and when there are 13 or more, the person is banned from driving

random breath test the power of the authorities, such as the police, to stop drivers at random (meaning, not needing to wait until they have committed some driving offense), and test them for breath alcohol

road user any person using the road, such as a driver, motorcyclist, bicyclist, pedestrian, jogger, etc.

sobriety test a simple chemical test that can be carried out to find out the level of alcohol in a person's breath, using a breathalyzer

unit of alcohol a convenient way to relate quantities of alcohol itself to the amounts of common drinks. One unit is the amount of alcohol contained in a conventional "single" drink, such as a small glass of wine, a shot of liquor or a bottle of ordinary strength beer or lager

INDEX

Action on Alcohol Abuse 60
Action on Drinking and Driving 60
age 33, 36, 39
Alcohol Concern 60
alcoholic 28
attitudes 6, 24, 26
Australia 22
Automobile Association 60

balance 30, 38
beer 34, 35, 36, 38, 39, 52, 53, 59, 61
behavior 6, 26, 29
BITER 60
blood 33, 38, 39, 46, 59, 61
blood test 38, 42, 61
body size 33, 39
brain 28, 29, 30, 33
breath 34, 38, 59, 61
breath test 21, 22, 34, 38, 46, 61
breathalyzer 38, 39, 42, 61
Britain 5, 6, 21, 24, 26, 38, 42, 47, 48

CADD 60
California 36
Canada 39
Chicago 49
cider 35, 59
coordination 6, 38, 59

depressant 29
disqualification 47, 48
driving license 24, 46
drug 28, 29, 34, 36

effects of alcohol 29, 34
England 10, 14, 20, 38, 47, 56
Essex 56
ethyl alcohol 61
experience 36, 52

gin 35
Glasgow 32

Health Education Authority 60

Idaho 38
Illinois 12
Institute of Alcohol Studies 60
insurance 48, 59
intoximeter 61

judgment 30, 38, 59, 61

lager 35, 52, 61
laws 6, 19, 21, 22, 24, 26, 41, 46, 47
legal limit 6, 26, 36, 38, 41, 42, 46, 48, 52, 61
liver 28, 34
Liverpool 52
London 22, 41, 50, 54

National Highway Traffic Safety Administration (NHTSA) 13
nervous system 30
New South Wales 22
New York 20
New Zealand 26
non-alcohol drinks 52, 53, 55

overdose 28
Oxford 10

penalties 46, 48, 60, 61
police 21, 22, 24, 38, 46, 58, 61
prison 10, 12, 48, 59
public transport 56, 57
punishment 12, 25, 47, 48

reaction times 30, 33
regulations 21
RoSPA 60

Scotland 32
sherry 35, 59
skill 19, 30
speed limit 58
stimulant 29
stomach 33, 34, 39, 59

United States 5, 6, 13, 15, 22, 26, 38, 48
units of alcohol 34, 35, 36, 39, 59, 61
urine 34, 38, 39, 59
Utah 38

whiskey 31, 35
wine 35, 36, 39, 52, 53, 54, 59

Photographic Credits:
Cover and pages 7, 17 and 28: Barry Lewis/Network; pages 4, 18, 41, 47, 49, 51, 53 and 58: Zefa; pages 9, 14, 15, 37 and 54: Robert Harding Library; pages 10 and 45: Hutchison Library; pages 11, 25, 55 and 58: J. Allan Cash Library; pages 23, 27, 32 and 57: Roger Vlitos; page 29: Mike Goldwater/Network; page 31: Spectrum Colour Library; page 34: Science Photo Library; pages 39, 42 and 47: The Metropolitan Police.